Earth Sc

CHALLENGE!

A CLASSROOM QUIZ GAME

Glen Phelan

WALCH PUBLISHING

1 2 3 4 5 6 7 8 9 10

ISBN 0-8251-5043-4

J. Weston Walch, Publisher
P.O. Box 658 • Portland, Maine 04104-0658
walch.com

Printed in the United States of America

Contents

To the Teacher

Earth Science Challenge! deals with facts and concepts about all aspects of the earth. It is designed to be used for several purposes: as a fun and easy way to reinforce what is being studied, as a study guide, and as a review of the unit or a culminating activity. It challenges your students to remember important facts and concepts and encourages them to enjoy themselves in the process.

The format of *Earth Science Challenge!* is similar to that of a popular television game show. A student is given the answer and is asked to provide the question. The fact given as a question is actually stated ("The blanket of gases that surround the planet"), not asked. The student response is given as a question ("What is the atmosphere?"). Many students will already be familiar with the format.

The questions are classified according to general topic and further by category. This format lends itself to use with a variety of attention-keeping games. Some games are suggested here; you and your students may invent others.

A few questions throughout the book have more than one correct response. Often, an alternate response is an abbreviation or an acronym for the full response, for example, "What is the U. S. Geological Survey (USGS)?" Other times, the alternate response is different but related, for example, "What is a cliff or steep slope?" In either case, the alternate response is given.

How to Use This Book

The arrangement of sections within each unit of *Earth Science Challenge!* closely follows the arrangement of topics in most popular earth science textbooks. This makes it easier to correlate the various games of *Earth Science Challenge!* with your course.

Each topic, or game, consists of five general categories. Within each category are five questions, each assigned a point value of 5 through 25, depending on its relative difficulty, plus a bonus question. The bonus question is not necessarily more difficult; it may refer to an unusual fact or a less important one. It may be used in whatever way seems suitable. A point value of 5 for each bonus question would give the entire game 400 points; a value of 25 would make it a 500-point game.

The questions in this format may be used to play a variety of games. However, it may prove effective to allow the students an opportunity to find the answers to, or study, the questions first. You may wish to reproduce the questions for a series of assignments, and then use a game as an evaluation, a further review, or a culmination of the unit. You may find that using the questions without a game is adequate. For these reasons, the answers are presented separately at the back of the book rather than with the questions.

Feel free to modify *Earth Science Challenge!* If you have stressed something in your class that is not included in this game, it is easy to add questions. Your students will quickly learn how to make questions for you in order to extend the game. You can also modify the questions to make them easier or harder to fit the needs of each particular class. Your class can play the same game more than once, which will help them remember material.

The same basic procedure can be used for playing any number of different games. Here are the directions for a typical game:

- On the board, write the categories for the game to be played along with point values for each question.
- Divide the class into teams. Play begins when one student asks for a question from a given category with a given point value. For instance, the student might say, "Formation of Sedimentary Rocks for 10 points."
- The game leader then reads the 10-point question from the category "Formation of Sedimentary Rocks."
- Any student on the team may answer. The first person on the team to raise his or her hand is called on. (It may be the student who asked for the category to begin with.)
- If the answer is correct, record points for the team. The student who answered chooses the category and point value for the next question.
- If the answer is wrong, subtract the point value of the question from the team score.

A student from the other team now has the chance to answer the question. Whoever answers the question correctly chooses the category and point value for the next question.

- If no one can answer the question, give the correct answer to the group. The student who last successfully answered a question chooses the next category and point value.
- When all the questions in the category have been used, erase the category from the board. Continue until all the categories are erased and the game is over.

Following are some other variations of the game:

Rounds

The categories and point values are displayed, and the value of the bonus question is agreed upon. Bonus questions are not used until last. A scoreboard is drawn on the board to show the teams and what score they receive in each round.

The class is divided into three, four, or five groups, each having an equal number of students. (Up to 30 can play. Extra pupils may serve as scorekeepers, readers, or board keepers.) The players in each group sit or stand in a set order—first player, second, and so forth.

The game begins with Player 1 on Team 1 requesting a question. If the player responds correctly, the earned score is recorded under Team 1/Round 1. If the response is incorrect, the correct answer is read and a score of zero is recorded. In either case, the point value is erased under the respective category. Then Player 1 of Team 2 has a turn to choose a question. After all the first players on each team have played, the play goes to the second players of each team, then the third, and so forth.

The game continues for as many complete rounds as possible. There may be several unused questions. If there are 30 players, the last player in each team chooses a category for a bonus question. Otherwise, the bonus question for each team is given to, or chosen by, the team's top scorer or chosen captain, either for that player or for the team to answer. The top-scoring team wins.

Progression

This game is set up like Rounds, preferably in five groups. The first players on each team choose a category for 5 points, the second players choose a question for 10 points, the third players go for 15, and so forth. Play continues for as many complete rounds as possible, with bonus questions handled as in Rounds.

Concentration

First, the categories and point values are written on the board, and the bonus value is determined. The class is divided into two teams. The first player on one team requests a question. If the player replies correctly, his or her team gets the points, and the point value is erased below the respective category. If the player does not answer correctly, the response is announced to be wrong and nothing is erased from the board. The first person on the opposite team then chooses a question. The play goes from team to team, with each person choosing a question still listed on the board. The advantage goes to the person who knows the answer to a previously asked question and can remember where it is located on the board. Play continues until all questions have been used. The highest-scoring team wins.

Last Chance

The class is divided into two, three, four, or five teams, with the players seated or standing in a set order. The categories and point values are displayed, and the bonus value (perhaps generous) is chosen. The bonus questions are not used in regular play.

Player 1 on the first team requests a question. If the player replies correctly, his or her team earns the respective points; if the reply is incorrect, the teacher tells or explains the answer. In either case, the point value under that category is erased. The play then goes to Player 1 on the second team, who requests a question. After all the first players have had a turn, the play goes to the second players on each team, then the third, and so forth.

When all the questions have been used, the scores for each team are calculated. The next player on the lowest-scoring team chooses a category for the bonus question for his or her team. The teacher reads the question and accepts only one answer from the team. (The players may confer in order to come to an agreement.) If the reply is correct, the bonus score is added to their total. Then the second-lowest-scoring team chooses a category, then the third, and the fourth, if there are that many teams. Only one bonus question is given to each team. There may be some that are not used. The winning team is the one with the highest score.

Solo

This game is played like Last Chance, except that it is played by five players instead of teams. The play goes from one player to the next in succession until all questions are used. Then each has a chance to choose a bonus question to raise his or her score. The top scorer wins.

Earth Science Bee

This game is played like a spelling bee, but no one is eliminated. First the categories and point values are displayed, and the value of the bonus question is determined. The class is divided into two teams. The first person on one team asks for a question by stating a category and point value. If the player responds correctly, his or her team receives the points, and that point value is erased under that category. The next turn is taken by the first player on the other team, who chooses a question. However, if the first player's response is not correct, the same question is repeated for the first player on the other team. If the player replies correctly, his or her team gets the points, and the play then goes to the second player of the first team. The play continues from one side to the other, with points going to the teams that answer correctly and the respective category points being erased from the board. The game is over when all 30 questions have been used. The team accumulating the most points wins.

No matter how you use *Earth Science Challenge!* it is an entertaining and stimulating way to review, and it's an excellent change-of-pace activity. You'll find your students eager to play it again and again.

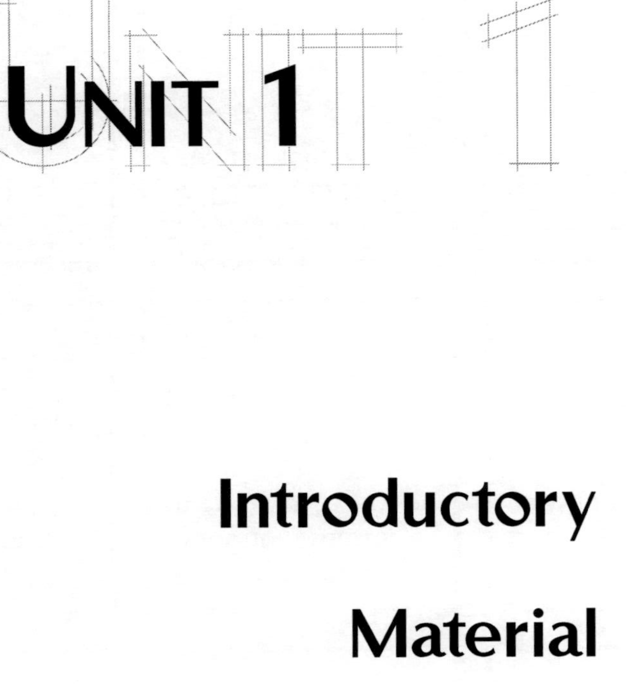

UNIT 1

Introductory Material

1 Introductory Material

Section 1

	Nature of Earth Science	Earth's Systems	Scientific Methods	Measurement I	Measurement II
5	The study of outer space and the objects in it	The blanket of gases that surround the planet	The first step in problem solving	The SI base unit of length	1,000 times greater than a meter
10	The study of the oceans	All the water that exists on Earth	A suggested explanation for an observation	A measure of the amount of matter in an object	The number of millimeters in a centimeter
15	The study of weather	All the organisms on Earth and the places they live	A standard for comparison in an experiment	A measure of the gravitational force on an object	The metric unit of temperature

20	The study of the materials that make up Earth	The hard outer shell of the planet	The factor that is manipulated in an experiment	A measure of how much space an object occupies	The measure of surface within set boundaries, expressed in square units
25	The application of scientific knowledge	The partially melted layer of the upper mantle	The explanation of facts that have been observed over a long period of time	Mass divided by volume	The SI unit of temperature
BONUS	The study of water as a resource	Nitrogen and oxygen	A statement of fact meant to explain observations unvarying over time and accepted as universal	The SI base unit for volume of solids	The modern version of the metric system, abbreviated SI
NOTES					

1 Introductory Material
Section 2

	Models	Parts of a Map	Types of and Projections of Maps	Topographic Maps I	Topographic Maps II
5	A representation of how something looks or works	Pictures or colors on a map that stand for features on Earth's surface	A map that uses lines to show the shape and elevation of the land surface	Connects points of equal elevation on a map	The type of scale (large or small) that most topographic maps are
10	A sphere that represents all of Earth's surface	A list that explains what all pictures and colors used on a map mean	A map that shows the distribution of different rock types at the surface	The change in elevation between two side-by-side contour lines	A spot on a map marked with BM and exact elevation
15	A representation on a flat surface of all or part of Earth's surface as seen from above	The ratio between distances on a map and distances on Earth's surface	A map projection in which areas near the poles are greatly distorted	Represented by hatchmarks, or hachures, pointing inward along a line	A series of closed contour lines, one inside the other, shows this land feature.

20	A model that represents an observable object at a different size	An item on a map that shows direction	This map projection is made by projecting points and lines from a globe onto a cone.	The agency in the United States that makes topographic maps	Contour lines that cross a stream point this way.
25	A model that represents something that cannot be seen	A large-scale map shows more of this than a small-scale map.	This map projection is often used for showing the polar regions and planning long-distance trips by air and sea.	Every fifth contour line	A series of contour lines very close together show this land feature.
BONUS	A model of the planets in motion around the Sun	A scale in which every unit on the map represents 63,500 units on Earth's surface	Mapmakers	Urban areas are this color on a topographic map.	Of these three, the contour interval most likely to be used on a map of mountainous terrain: 10, 30, 100
NOTES					

1

Introductory Material

Section 3

	Latitude and Longitude	Time Zones	Remote Sensing	Continents	Landforms
5	Degrees of this tell you how far north or south of the equator a place is	Earth rotates in this direction.	Objects that orbit and collect information about Earth	The number of continents	Landforms that reach the highest elevations
10	Degrees of this tell you how far east or west of the prime meridian a place is	The number of time zones on Earth	A type of satellite that maps Earth	The combination of Europe and Asia	Low areas between hills or mountains, often containing streams
15	Any place north of the equator is in this part of the world.	While traveling west across each time zone in the United States, time is affected in this way.	A radio-navigation system that consists of 24 satellites to determine position	The smallest continent	Broad, flat areas of land that are not too high above sea level

20	A degree is divided into 60 of these.	The line on Earth at which the date changes	The use of sound waves to map the ocean floor	The continent that is almost completely covered with ice	Broad, flat areas of land that are well above sea level
25	The latitude of the South Pole	When it is 9:00 A.M. in San Francisco, it is this time in New York City.	Computers use satellite data to produce these pictures of Earth that are not true-to-life colors.	Two continents connected by the Isthmus of Panama	Low, flat areas of land along the coast
BONUS	This line of longitude goes through Greenwich, England.	The width, in degrees, of each time zone	The part of a GPS system that someone on Earth uses	The two continents through which the equator runs	The difference in a region's elevations
NOTES					

UNIT 2

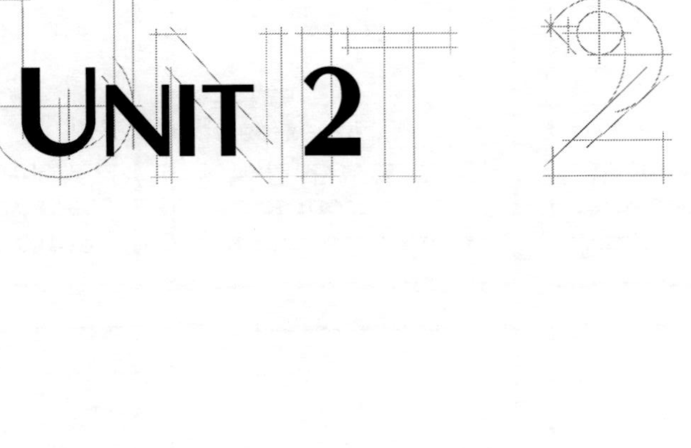

Earth Materials

2

Earth Materials
Section 4

	Atoms	Phases of Matter	Elements	Combinations of Matter	Energy
5	A substance that cannot be broken down into simpler substances	The phase in which a substance has a definite shape and volume	The atoms of an element all contain the same number of these	A substance made of atoms of two or more elements	Energy of motion
10	The smallest particle of an element that has the characteristics of that element	The phase in which a substance has a definite volume but not a definite shape	The number of protons in an atom's nucleus	Atoms that share electrons are held together by this force.	Stored energy
15	A particle in an atom's nucleus that has a positive charge	The process of changing from a liquid to a gas	Atoms of the same element that have different numbers of neutrons	A mixture in which one or more substances is dissolved in another	A form of energy released when atomic nuclei split or combine

20	A particle in an atom's nucleus that has no charge	The process of changing from a gas to a liquid	A table that arranges elements by their chemical properties and atomic number	The smallest unit of a compound in which the atoms are held together by sharing electrons	The sun's energy comes from the combining of these.
25	An atom that gains or loses an electron	Hot gas made of positive ions and free electrons	The total number of protons and neutrons of an atom	A solution of two or more metals	The statement that energy is conserved in any reaction
BONUS	The number of elements that occur naturally	The process of a solid changing directly to a gas	The most abundant element in Earth's crust	A molecule composed of two hydrogen atoms and one oxygen atom	Chemical energy results when these are broken or made.
NOTES					

2 Earth Materials

Section 5

	Mineral Characteristics	Mineral Groups	Identifying Minerals I	Identifying Minerals II	Mineral Resources
5	What all minerals are, meaning they are not alive and never were alive	The most common mineral group, these minerals are made largely of oxygen and silicon.	A few minerals, such as sulfur and malachite, often can be identified by this property.	A mineral that breaks along one or more smooth planes has this.	A mineral or group of minerals from which a useful material can be extracted at a profit
10	How all minerals form, as opposed to synthetically	A common silicate, it forms most of the sand on beaches.	The way a mineral reflects light	A mineral that breaks with a rough or jagged edge has this.	Minerals prized for their rarity and beauty
15	The phase of all minerals	The most abundant silicate, it weathers into clay.	Examples include smooth, rough, greasy, and glassy.	Carbonate minerals give off this when in contact with acid.	A band of minerals that forms in a crack in surrounding rock

20	A solid in which the atoms are arranged in a repeating pattern	These minerals are composed of one or more metals with the compound CO_3.	The color of a mineral when it is powdered	Calcite has this property of splitting light rays into two parts.	A type of mine in which overlying rock is removed, creating a huge hole
25	The material that cools to form many minerals	These minerals are compounds of oxygen and a metal.	A measure of how easily a mineral can be scratched	The comparison of a mineral's density with the density of water	The main ore of aluminum
BONUS	Within 500, the general number of known minerals	The basic building block in silicates, consisting of four oxygen atoms bonded to a central silicon atom	The hardest mineral known	This mineral is magnetic, sometimes called lodestone.	A gem of quartz, made purple by trace elements

NOTES

2 Earth Materials

Section 6

	Introduction to Rocks	Formation of Igneous Rocks	Classifying Igneous Rocks	Formation of Metamorphic Rocks	Classifying Metamorphic Rocks
5	Rocks are made of one or more of these.	Molten material that cools and hardens underground to form igneous rocks	Igneous rocks are classified by texture and this property.	The general name of the existing rock that gets changed into metamorphic rock	The two groups into which metamorphic rocks are classified
10	A repeated series of events that changes rocks from one form to another	Molten material that cools and hardens on Earth's surface	This light-colored intrusive igneous rock contains quartz and feldspar.	A type of metamorphism that occurs over a large area	Metamorphic rocks that have visible layers or bands
15	A type of rock formed from the cooling and hardening of molten rock	Igneous rocks that formed inside Earth	This lightweight rock forms from lava and resembles a sponge.	A type of metamorphism that occurs when magma comes in contact with rock	A metamorphic rock formed from shale, it splits easily into sheets.

20	A type of rock formed from the compacting and cementing of sediments	Igneous rocks that formed on Earth's surface	The most common rock of lava flows	A type of metamorphism that occurs from the stress of rocks moving past each other at faults	A metamorphic rock formed from limestone
25	A type of rock formed by other rocks that are under extreme heat and pressure	The texture that igneous rocks have when they form so quickly that crystals don't have time to form	An igneous rock in which large crystals are surrounded by fine-grained crystals	A type of metamorphism that occurs when hot water comes in contact with rock	A metamorphic rock formed from granite
BONUS	Igneous comes from the Latin *ignis,* which means this.	Magma that is thick and slow-moving is felsic. Magma that is thinner and more fluid is this.	The extrusive counterpart of granite	Large crystals that form in solid rock during metamorphism	Metamorphosed sandstone
NOTES					

2

Earth Materials

Section 7

	Formation of Sedimentary Rocks	Classifying Sedimentary Rocks I	Classifying Sedimentary Rocks II	Features of Sedimentary Rocks	Name That Rock
5	Small, solid particles that form sedimentary rock	A type of sedimentary rock formed from fragments of other rocks	Clastic rock made of rounded pebbles	Evidence of past life often found in sedimentary rocks	Igneous rock of interlocking white, pink, gray, and black crystals
10	The process of sediment being deposited	A type of sedimentary rock formed from seashells and other remains of organisms	Clastic rock made of silt and clay-sized particles	Horizontal layering of sedimentary rocks	Igneous rock that is dark and glassy, and fractures with curved surfaces
15	The process of sediment being pressed together	A type of sedimentary rock formed from the precipitation of dissolved minerals	An organic rock made of microscopic shells; useful in school	Small ridges in sediment formed by wind, waves, or currents	Coral reefs will eventually become this kind of sedimentary rock.

20	The process in which minerals fill the spaces between sediments, binding them together	The most abundant organic sedimentary rock, composed mostly of calcite	Clastic rock made of quartz sand grains	Inclined layers of sediment that move across a horizontal surface	Metamorphic rock sometimes used as roofing tile
25	The physical and chemical processes that change sediments into sedimentary rock	Clastic rocks with sand-sized grains are medium-grained; clastic rocks with gravel-sized grains are this.	This rock will fizz when dilute hydrochloric acid is placed on it.	A sphere of rock formed when crystals grow in a cavity in limestone	Like a conglomerate except its particles are sharp and angular
BONUS	Rocks with a cementing material of silica or calcite tend to be this color.	An organic rock commonly used for fuel	An organic rock made from mostly whole seashells	A round lump of calcium carbonate that forms as minerals precipitate around a shell fragment or sand grain	Metamorphosed basalt

NOTES

UNIT 3

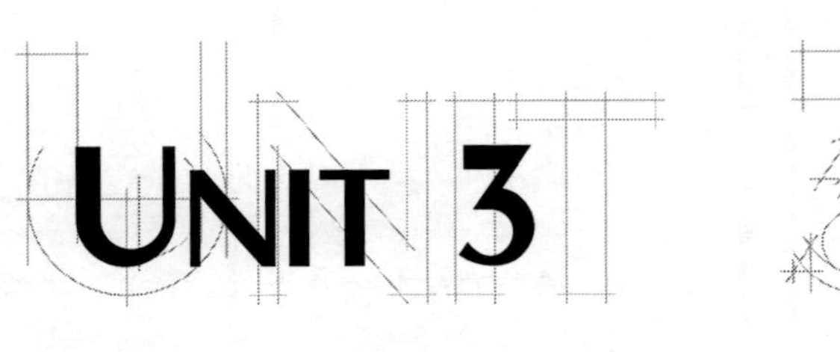

Earth's Internal Processes

3 Earth's Internal Processes

Section 8

	Structure of Earth	Drifting Continents	Seafloor Spreading	Plate Tectonics I	Plate Tectonics II
5	The thin, rigid outer layer of Earth that includes the surface	The hypothesis that Earth's continents were once joined in a single landmass	A long narrow chain of mountains that runs along the Atlantic Ocean floor	A slab of lithosphere that moves on Earth's surface	These flows of energy in the upper mantle may cause plate motion.
10	The thickest layer of Earth, mostly solid	The name of the supercontinent formed by the combining of Earth's landmasses	The deepest regions of the ocean floor	A place where two plates are moving toward each other	Hot mantle material rises because of this property, compared to its surroundings.
15	The innermost part of Earth's center, composed of solid iron and nickel	Evidence that continents were combined includes the puzzlelike fit of Africa and this continent.	Compared with rocks near the trenches, rocks near the ocean ridges are this.	A place where two plates are moving away from each other	A force that pushes an oceanic plate toward a trench

20	The outermost part of Earth's center, composed of liquid iron and nickel	Evidence that continents were combined includes signs of once-living things called this.	This is formed at ocean ridges.	A place where two plates are sliding past each other	A force in which the weight of a subducting plate pulls on the rest of the plate
25	The crust and uppermost part of the mantle that is rigid	The hypothesis of continental drift was rejected for a long time because it could not explain this.	Rocks on the ocean floor show reversals in the polarity of this over geologic time.	The process of one plate descending beneath the other	Pangaea first broke up into two large landmasses: Laurasia and this.
BONUS	The slushy layer of the upper mantle upon which the lithosphere floats	A scientist who proposed the hypothesis of continental drift	A scientist who proposed the theory of seafloor spreading	Within 2 cm per year, the rate at which the Atlantic Ocean is growing	The name of a valley in East Africa where two plates are spreading
NOTES					

3 Earth's Internal Processes

Section 9

	Faults	Earthquake Waves	Measuring and Locating Earthquakes	Earthquake Hazards	Earthquake Prediction
5	A break in rock along which movement occurs	Energy moves away from an earthquake as these kinds of waves.	The point at which rock first breaks or moves to cause an earthquake	The most damaging earthquakes have this type of focus.	One of the three states with the highest risk of earthquakes
10	A fault in which the rocks on either side move horizontally past each other	The fastest seismic waves, these can travel through liquids or solids.	The point on the surface above where an earthquake originates	Huge ocean wave caused by underwater earthquakes	Earthquake damage in high-risk areas can be minimized by not building on or near one of these.
15	A normal fault forms when this moves down compared to the footwall.	Seismic waves that cause rocks to move at right angles to the wave direction	An instrument that measures waves produced by earthquakes	Smaller earthquakes that follow a larger one	A worldwide map of earthquake epicenters closely resembles a worldwide map of these.

20	A fault caused by compression of rock	Seismic waves that cause rocks to move up and down, and side to side	Common scale used to describe an earthquake's magnitude, or amount of energy released	In addition to collapsing buildings, the other major hazard of earthquakes in populated areas	A section of an active fault that has not had an earthquake for a long time
25	A reverse fault in which the fault plane dips 45 degrees or less from the horizontal	Of the three types of seismic waves, these cause the most damage.	Minimum number of seismograph stations needed to find an earthquake's epicenter	The process in which severe ground shaking causes soil to act like a liquid	This may be building up along a seismic gap.
BONUS	A cliff formed along a fault line	P waves and S waves are also called these.	The most accurate scale that scientists use to measure an earthquake's magnitude	Measures earthquake intensity based on the amount of damage caused	To be most effective, an earthquake prediction should include time, location, and this.
NOTES					

3 Earth's Internal Processes

Section 10

	Magma	Volcanic Landforms	Volcanoes and Plate Tectonics	Intrusive Features	Volcanic Eruptions
5	Magma that reaches Earth's surface	The opening in the crust through which lava erupts	Most volcanoes form at these.	A general term for an intrusive body of rock	A volcano that erupted in Washington state in 1980
10	A magma's resistance to flow	A type of small volcano with steep sides	A chain of volcanic islands formed where two oceanic plates collide	An intrusion of magma that cuts across preexisting rocks	The rapid movement of ash, gas, and other materials down the outside of a volcano
15	This magma formed the basalt rock of the Columbia Plateau in the northwestern United States.	The Hawaiian Islands are formed from this type of volcano.	Unusually hot region in Earth's mantle where plumes of hot material rise	The largest plutons	Lava flow that hardens into rock with smooth, ropelike surfaces

20	This magma fuels volcanoes that erupt explosively.	A large explosive volcano formed of alternating layers of volcanic fragments and lava	An island in the North Atlantic that has several volcanoes due to seafloor spreading	A pluton that forms when magma intrudes between parallel layers of rock	Lava that cools underwater and looks like smooth lumps
25	An increase in temperature, a decrease in pressure, or an increase in water in the asthenosphere	The result of many episodes of basaltic lava pouring onto land from a fissure	An area of fractures and faults through which magma rises at divergent plate boundaries	Similar to but smaller than a batholith	An instrument placed on a volcano that detects changes in the slope of the volcano
BONUS	This magma has intermediate viscosity, gas content, and silica content.	A large basin that forms when the top of a volcano collapses	Indicated by the bend in the Hawaiian-Emperor volcanic chain	A ring-shaped island that forms when an oceanic volcano collapses or erodes	Fragments of rock thrown into the air during a volcanic eruption
NOTES					

3

Earth's Internal Processes

Section 11

	Isostasy	Convergent Boundary Mountains I	Convergent Boundary Mountains II	Divergent and Nonboundary Mountains	Mountain Ranges
5	The crust floats on the mantle because it has less of this property than the mantle.	Convergent boundary mountains form when these collide.	The tallest mountains form when these two kinds of plates collide.	A mountain of molten rock and ash that builds around an opening	The tallest mountain range in the world
10	The upward force on crust that balances the downward force of gravity	Mountains formed when rock layers are squeezed from opposite sides	Major mountain belts are NOT formed when these two kinds of plates collide.	Mountains that form when crust is pushed up from below	A mountain range in the eastern United States that formed at an ancient plate boundary
15	The principle that describes the displacement of the mantle by Earth's crust	A downfold in rock layers	Major mountain belts near a coast are formed when these two kinds of plates collide.	Mountains formed from blocks of crust that have been faulted and tilted	This process has made the Appalachians lower than the Rockies.

20	The crust with greater density: oceanic or continental	An upfold in rock layers	The two sides of a fold are called these.	Mountains formed at underwater divergent boundaries	Uplifted mountains in South Dakota
25	The rising of the crust as erosion removes mass	Stress that involves squeezing	The cycle of processes that form all mountain ranges	Stress that involves pulling and stretching	Folded mountains formed from the collision of the Nazca and South American plates
BONUS	The crust that projects into the mantle beneath mountain ranges	The direction of a fold of rock	The steepness of the limbs of a fold	A block of crust thrust upward between two normal faults	A fault-block mountain range in Wyoming
NOTES					

UNIT 4

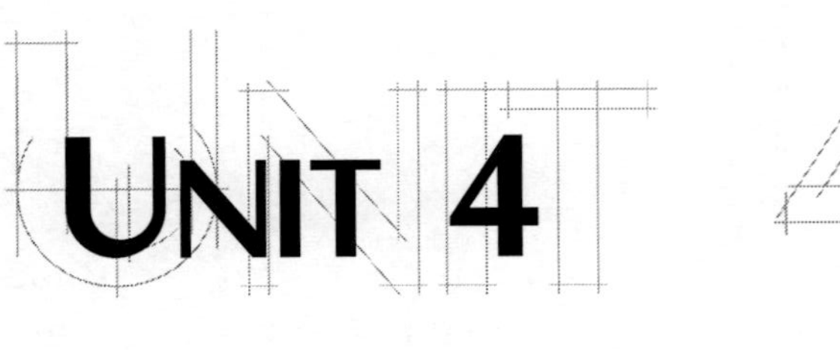

Earth's Changing Surface

4 Earth's Changing Surface
Section 12

	Mechanical Weathering	Chemical Weathering	Weathering Rates	Soil Formation	Soil Types
5	The process that breaks down and changes rocks at the surface	The common name for iron oxide that forms when iron weathers	Chemical weathering occurs faster in this kind of climate.	Decayed organic material in soil	Soils are classified mostly based on this.
10	Mechanical weathering changes a rock's size and shape but not this.	The reaction of water with other substances	Multiple cracks in rock expose more of this to weathering.	The main component of soil	Smallest soil particle
15	Plant parts that can split rock as they grow	The chemical reaction of oxygen with other substances	Of the three major rock types, the one most easily weathered	A vertical sequence of soil layers	Soil composed of a combination of clay, silt, and sand

20	The repeated freezing and thawing of water in cracks of rock	A weak acid formed when carbon dioxide dissolves in rainwater	Of quartz, calcite, and gypsum, the one most difficult to weather	A distinct layer of soil	A type of soil in cold climates at high latitudes and high elevations
25	The peeling of outer rock layers as pressure on rock is reduced	The mineral common in limestone that dissolves in carbonic acid	Little or no chemical weathering occurs in this kind of climate.	A layer of soil that usually includes much clay	A type of soil that covers most of the East Coast and Midwest in the United States
BONUS	Animals that do this in the soil create holes for air and water to reach bedrock.	Feldspar chemically weathers into this clay mineral.	Weathering occurs faster in this kind of topography.	A layer of soil that includes partially weathered parent material	A type of soil with little or no organic matter and a very thin A horizon
NOTES					

4 Earth's Changing Surface

Section 13

	River Systems	River Valleys	River Erosion	River Deposition	Floods and Floodplains
5	All rivers flow in this direction.	The cross-sectional shape of a youthful river valley	The removal and transport of Earth's material by water, wind, glaciers, and gravity	The dropping of materials carried during erosion	The part of a valley floor that gets covered with floodwaters
10	A river that runs into another river	A river valley with steep vertical sides	The process of erosion in which solid particles scrape against rock and wear it away	A fan-shaped deposit of sediment at the mouth of a river	A meander that's been cut off from the river
15	All the land whose water drains into a certain river system	A large groove or channel in soil that carries runoff after rain falls	Erosion occurs mostly along this part of a meander.	Deposition occurs mostly along this part of a meander.	Erosion occurs mostly in this direction on a floodplain.

20	The steepness of a river's slope	A looping bend in the course of a river	The wearing away of land at the head of a gully or river valley	A fan-shaped, sloping deposit of sediment that forms where a river leaves a mountain range	A ridge built up along the banks of a river, either natural or artificial
25	The high ground between two drainage basins	The stage during which a river starts to erode its valley sideways as well as down	The eroded material carried downstream	The branches of streams that form on a delta	A sudden flood, usually caused by heavy rainfall
BONUS	The treelike pattern that a river system forms	The elevation at which a river enters another river or body of water	The movement of water in all different ways instead of straight downstream	A major city located on the Mississippi River delta	A tributary that forms on the floodplain parallel to the main river for a distance
NOTES					

4 Earth's Changing Surface

Section 14

	Groundwater	Groundwater Erosion and Deposition	Mass Movements	Wind Erosion	Wind Deposition
5	The highest level underground that is saturated with water	Caves often form when groundwater does this to limestone.	The force that moves rock and soil downhill	A storm with strong, steady winds that blow great amounts of soil	A hill of sand deposited by wind
10	A small stream of water that forms where the water table meets the surface on a hillside	A depression formed when the roof of a cave collapses	The rapid flow of mud and water down a hillside	Wind erosion is most effective in this climate.	The wind-blown deposits of silt that accumulate in thick layers
15	The underground layer where all the pores are filled with groundwater	A type of landscape consisting of caves, sinkholes, and valleys in limestone	The slow movement of material downhill that causes objects to tilt	The removal of loose rock particles by wind, lowering the land surface	The side of the dune that faces the wind, this usually has the gentler slope.

20	The ability of rock to let water pass through its pore spaces	Mineral deposits that hang like an icicle from a cave ceiling	A mass of bedrock or loose soil or rock downhill	A surface made of pebbles and boulders after wind has blown away sand and silt	The side of the dune that usually has the steeper slope
25	The pores in this area above the water table contain mostly air.	Mineral deposits that grow upward like a cone from a cave floor	The downhill movement of a block of land along a curved surface	The method of wind transport in which sand grains bounce along the surface	The mineral that makes up most of the sand in dunes
BONUS	The process by which water rises slightly because its molecules are attracted to soil particles	Water that contains high concentrations of calcium, magnesium, or iron is called this.	Fragments of rock that gather at the bottom of a slope	Rocks shaped by wind-blown sediments	The movement of a dune as wind blows the sand up and over the dune
NOTES					

4 Earth's Changing Surface

Section 15

	Glaciers	Glacial Movement	Glacial Erosion	Glacial Deposition	Ice Ages
5	A glacier is a large, moving mass of this.	The force that causes glaciers to move	The cross-sectional shape of a valley carved by a glacier	Unsorted sediments deposited by glaciers	These kinds of glaciers covered large parts of Earth's surface during an ice age
10	A glacier that forms in valleys in mountainous areas	The force that slows the movement of glaciers along its bases and sides	A semicircular basin carved out by a glacier at the head of a valley	A ridge of glacial till	These large lakes were formed during the last ice age.
15	A glacier that covers a large part of a continent	Large blocks of ice that break off glaciers and drift in the sea	A lake made from a chunk of glacial ice that makes a depression	Long, smooth hills formed as glaciers move over till	A large, hairy relative of the elephant that roamed North America during the ice age

20	The lowest elevation at which snow occurs year-round	Large cracks that form in the glacier as it moves over a steep slope	Three or more cirques together form this pyramid-shaped peak.	Material deposited by glacial meltwater	Within 3,000 years, how many years ago the ice sheets retreated
25	One of the two places that consist of the largest continental glaciers	One of the areas of a glacier that moves the fastest	An arm of the sea that fills a glacial valley along the coast	Long, winding ridge of sediments deposited by a stream beneath a melting glacier	The warm period between glacial advances in an ice age
BONUS	Compacted snow that is not yet glacial ice	Movement in which the glacier's base slides due to a layer of water that forms	Long parallel scratches left on bedrock after a glacier passes over	A boulder that has been transported by a glacier	Changes in Earth's landmasses may have contributed to ice ages by changing the patterns of these.
NOTES					

4 Earth's Changing Surface

Section 16

	WAVES	SHORELINE WAVES AND CURRENTS	SHORELINE EROSION	SHORELINE DEPOSITION I	SHORELINE DEPOSITION II
5	The highest point of a wave	The collapsing wave that forms when a wave reaches shallow water	A hollow area eroded out of the rock along the shore	An area of sand or other sediment along the coast that slopes to the water	A long ridge of sand deposited by longshore currents and separated from the mainland
10	The lowest point of a wave	This characteristic of a wave increases as it approaches shore.	A curved feature that forms when waves erode softer rock that underlies harder rock	Most of the sand on beaches is brought to the coast by these	These help to stabilize dunes along beaches.
15	The vertical distance between a wave's crest and trough	The water that flows back toward the ocean after breaking on shore	Along an irregular coastline, erosion tends to smooth out these areas.	The process that moves beach sediment along the beach	A curved stretch of land at the end of a spit

20	The horizontal distance from one crest to the next	A current that moves large amounts of sediment along the shore	The tower of rock that remains when a sea arch collapses	A narrow stretch of land that extends into the water from the shore	Walls of concrete or boulders built out into the water to trap beach sand
25	The time it takes one wavelength to pass a given point	The bending of waves as they reach shallow water	Waves carrying sediment erode the coastline through this process.	A spit that forms completely across a bay	Walls of concrete or boulders built in the water parallel to shore to protect the shore from waves
BONUS	The length of open water over which the wind blows	A strong surface current that flows away from the shore	A flat erosional surface of land at the base of a sea cliff	The area of water behind a baymouth bar	A ridge of sand that connects the mainland to an island
NOTES					

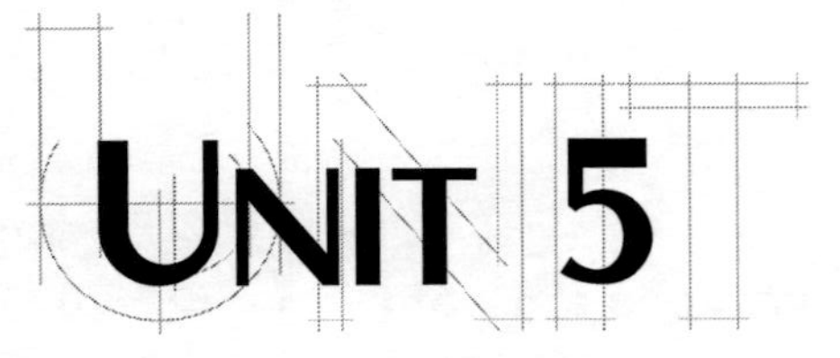

UNIT 5

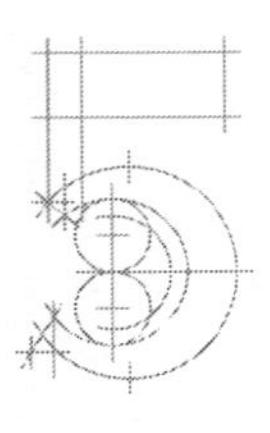

The Atmosphere

5 The Atmosphere

Section 17

	Composition of the Atmosphere	Structure of the Atmosphere	Heat in the Atmosphere	Air Pressure	Wind
5	About 99% of air is composed of nitrogen and this gas.	The atmospheric layer closest to Earth's surface	The source of the heat energy that enters the atmosphere from outer space	As elevation increases, air pressure does this.	Wind is air that blows from areas of high pressure to areas of this.
10	The amount of this gas in the air varies depending on the humidity.	In the troposphere, temperature does this the higher you go.	The transfer of energy through space by visible light and other electromagnetic waves	An instrument used to measure air pressure	Global winds that usually blow from the same direction
15	Air contains tiny particles of dust and this particle from ocean spray.	Ozone is present in this layer, above the troposphere.	The transfer of energy that occurs when molecules or atoms collide	As temperature decreases, air pressure does this.	A local wind that blows from the ocean toward the shore

20	A gas in the air formed by the addition of an oxygen atom to an oxygen molecule	The boundary between the stratosphere and the next highest layer	The transfer of energy through rising and sinking currents	A line on a weather map that joins points of equal air pressure	Narrow bands of high-altitude, high-speed winds
25	Besides nitrogen and oxygen, the remaining 1% of air is made of argon and this.	Part of the thermosphere, made of electrically charged particles	The measure of the average kinetic energy of the atoms or molecules in a substance	A unit of measure for air pressure used on weather maps	The effect of Earth's rotation that causes wind to deflect to the right or the left
BONUS	Within 10 percentage points, the percentage of oxygen in the air.	The peak of Earth's tallest mountain is in this layer.	The energy transfer that occurs because of a difference in temperature between substances	An Italian scientist who invented the mercury barometer	A zone of low pressure near the equator where air rises
NOTES					

5 The Atmosphere

Section 18

	Water Cycle	Humidity	Cloud Formation	Cloud Types	Precipitation
5	The water cycle circulates water between Earth's surface and this.	Humidity is a measure of how much of this is in the air.	For clouds to form, warm, moist air must rise and cool to this point.	Clouds on the ground	Cloud droplets may become large enough to fall as rain when they do this.
10	The process of liquid water changing to a gas	Air that contains all the water it can hold is said to be this.	Small particles in the air around which water vapor condenses	Fluffy clouds with flat bases	An instrument used for measuring rainfall
15	The process of water vapor changing into liquid water	The ratio of water vapor in air compared to how much water vapor that air can hold	When air warms and expands, its molecules do this, making the air less dense and causing them to rise.	High, feathery clouds made of ice crystals	Lumps of ice formed by layers of water freezing around frozen raindrops or ice crystals

20	Rain, snow, and other forms of water that fall to the ground	An instrument used to measure humidity	This is released when condensation occurs.	Low, layered clouds	Raindrops that freeze instantly when they hit a solid surface
25	The evaporation of water through leaves	The relative humidity of air that is holding 55% of all the water vapor it could hold at that temperature and pressure	Most tall clouds consist of water droplets and these.	These clouds combine the Latin words for "heap" and "rainstorm."	Particles of ice that form when raindrops freeze before reaching the ground
BONUS	The flow of rain over the ground and into a body of water	The two kinds of thermometers used in a psychrometer	The ability of air to resist rising	A prefix used to describe clouds at middle altitudes.	The process of cloud droplets joining to form larger droplets
NOTES					

5

The Atmosphere

Section 19

	Air Masses	Fronts	Pressure Systems	Collecting Weather Data	Forecasting Weather
5	These two characteristics are nearly uniform throughout an air mass.	A front is a boundary between two of these.	The Coriolis effect makes rising or sinking air do this.	A vehicle most often used to carry instruments high into the atmosphere	A scientist who studies and predicts the weather
10	Name of the cold, dry air masses from Canada	A front where cold air displaces warm air and forces it up steeply	The way that a high-pressure system rotates in the Northern Hemisphere	A type of radar that shows the movement of wind within a storm	Weather associated with low-pressure systems
15	Characteristics of maritime tropical air masses	A front where warm air displaces cold air and glides up over the cold air gradually	The way that a low-pressure system rotates in the Northern Hemisphere	An instrument used to measure wind speed	Weather associated with high-pressure systems

20	The name of air masses designated mP	A front that is not moving	Air flowing inward toward the center and upward describes this kind of system.	A property used by infrared satellite imagery to make images	Computer programs used to make weather predictions
25	The name of air masses that originate over deserts	A front that results when a cold front overtakes a warm front	The boundary between cold and warm air along which mid-latitude lows travel	A package of instruments carried by a balloon to measure the atmosphere	A government organization that produces weather forecasts
BONUS	The process of an air mass taking on the characteristics of a new region over which it travels	The first signs of the approach of a warm front are these.	Controls a low's path and strength	A computerized system that collects weather data from surface stations in the United States	A type of weather forecast that compares current weather patterns to those in the past
NOTES					

5

The Atmosphere

Section 20

	Thunderstorms	Tornadoes	Hurricanes	Other Extreme Weather	Weather Maps
5	The sound produced when lightning superheats the air suddenly	The kind of air pressure in the center of a tornado	An area of Earth where hurricanes begin to form	A winter storm of high winds, low temperatures, and falling or blowing snow	Represented by the letter H
10	The first stage of a thunderstorm when updrafts build cumulus clouds	A tornado begins when wind speed and this change between lower and upper winds.	A hurricane requires a steady supply of this.	A long period of little or no precipitation	Represented by a line with triangles pointing in the same direction
15	In the mature stage, precipitation creates downward winds called these.	The times of the year when tornadoes are most likely to occur	The center of a hurricane where the weather is often fair	The temperature that your body feels due to the combination of cold air and wind	Lines connecting points of equal air pressure

20	A very powerful single-cell thunderstorm	The center of a tornado	The high-level waters pushed onshore by hurricane winds	A sudden rush of water or overspilling of riverbanks usually caused by intense rain	Lines connecting points of equal temperature
25	Thunderstorms that form along cold or warm fronts	What the National Weather Service issues if conditions are right for a tornado to form	The location of the strongest winds in a hurricane	The temperature that your body feels due to the combination of hot air and humidity	A record of weather data for a certain place and time using weather symbols
BONUS	Most particles near the bottom of a storm cloud have this charge.	The name of the scientist for whom the tornado intensity scale is named	The minimum speed of sustained winds for a tropical storm to be called a hurricane	A winter storm along the eastern U.S. coast when winds blow from the northeast	Forecasters use temperature, wind direction, and this to locate fronts on maps.
NOTES					

5 The Atmosphere

Section 21

	Climate	Climate Classification	Natural Climate Changes	Humans Affect Climates	Seasons
5	The two main conditions that describe climate	The divisions of a major climate zone are called this.	When blocked by volcanic dust, this leads to a cooling of climates	The rise in worldwide temperatures possibly from CO_2 being added to the atmosphere	The cause of the seasons
10	The windward side of a mountain range is often wetter than this side.	Of the two polar subclimates, this one covers northern Alaska.	A warm ocean current in the Pacific Ocean that causes short-term climate changes	A scientist who studies climates	When it is summer in the Southern Hemisphere, it is this in the Northern Hemisphere.
15	Latitude affects climate because Earth is this.	The two dry subclimates are desert and this.	A variation in the tilt of this may cause climate changes.	The main activity of humans that adds CO_2 to the atmosphere	The name of the first day of summer in the Northern Hemisphere

20	The difference between the highest and lowest temperatures of the day	A localized climate that differs from the main climate of the region	Warmer climate changes correspond to an increase in this type of solar activity.	The heating of Earth's surface caused by certain atmospheric gases that absorb solar energy	The name of the first day of spring in the Northern Hemisphere
25	Coastal areas have milder climates because water does these two things more slowly than land.	An urban area where the climate is warmer than in the surrounding area	Past volcanic activity may have changed climates by producing large amounts of this gas.	This mass removal of trees increases CO_2 in the atmosphere.	The number of hours of daylight every place on Earth would experience if the axis were not tilted
BONUS	The warm ocean current that helps keep the climate of the United States East Coast mild	The name of the main system used to classify climates	A type of sample taken to study how ice records changed in past climates	In the greenhouse effect, absorbed solar energy is reradiated as this.	The line on the globe that marks the farthest point north that the sun ever appears directly overhead
NOTES					

UNIT 6

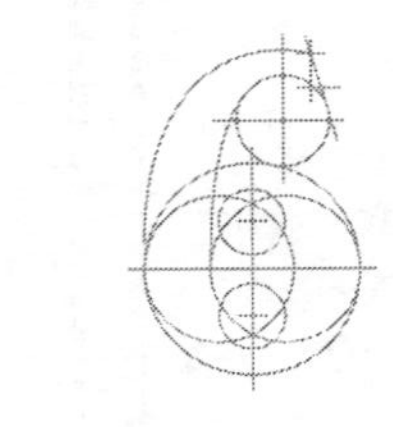

The Oceans

6 The Oceans

Section 22

	Oceans of the World	Properties of Seawater	Structure of the Oceans	The Ocean Floor I	The Ocean Floor II
5	The scientific study of the oceans	A measure of the dissolved salts in seawater	Water absorbs light, so below 100 meters, oceans are this.	A flat or gently sloping submerged part of a continental margin	A line of underwater mountain ranges with a total length of 65,000 km
10	The largest ocean in the world	The most abundant salt in seawater	How water temperature changes with depth	The sloping region at the edge of a continental shelf	Small, rolling hills on the ocean floor
15	The smallest ocean in the world	The highest seawater salinities occur where rates of precipitation are exceeded by rates of this.	Wind and waves mix heat evenly throughout this ocean surface region.	The long, narrow, steep-sided troughs on the ocean floor at subduction zones	An underwater canyon that cuts through the continental shelf and slope

20	A technique of using reflected sound signals to measure water depth	Of seawater and fresh water, the density of this is greater.	The region of ocean just below the mixed layer where temperature decreases rapidly	The smooth, flat part of the seafloor	A hole in the seafloor through which heated fluid erupts
25	Within 5%, the percentage of Earth covered by oceans	Of seawater and fresh water, the freezing point of this is higher.	Most fish and other free-swimming marine organisms live in this layer.	An underwater volcano that does not rise above the ocean surface	A sloping accumulation of sediments at the base of a continental slope
BONUS	A British ship on which the first large-scale research of the ocean was conducted in the 1870s	The average salinity of seawater	Dense water mass found beneath almost all other ocean water	Flat-topped seamounts	An area along an ocean ridge that is broken into stepped sections by transform faults
NOTES					

6 The Oceans
Section 23

	Surface Currents	Underwater Currents	Tides	Ocean Life	Ocean Resources
5	The main cause of surface currents	The movement of water that sinks and flows beneath water that is less dense	The force of attraction that causes tides	Microscopic plants that float freely with the ocean currents	A common resource obtained by letting shallow pools of seawater evaporate
10	Currents flowing toward this imaginary line carry cold water.	Seawater becomes more dense when it has more sediment, is saltier, or is this.	The body in space that most affects tides	Free-swimming organisms	The process of removing the salts from seawater to make fresh water
15	A current that flows opposite the wind-related current	The movement of cold, deep water coming to the surface	A tide that occurs when the Sun and Moon are at right angles	Microscopic animals that float freely with the ocean currents	Large seaweed used as a food source

20	The direction ocean currents turn in the Southern Hemisphere	A rapid current that carries mud and sand like an underwater landslide	A tide that occurs when the Sun, Moon, and Earth line up	Tiny, colonial sea animals that form many reefs	The process of removing too many fish from an area to sustain their population
25	Most surface currents are caused by these two types of prevailing winds.	These are contained in large amounts in upwelled water, which helps marine animals thrive	The difference in water level between high and low tide	The process by which bacteria near deep-sea vents use chemicals to produce food	The main metal contained in nodules on the seafloor
BONUS	The loop formed by a closed circular current system	Upwelling occurs mainly on this side of continents.	The location of the greatest tidal range on Earth	The gas used by bacteria for food near deep-sea vents	Sediments on the ocean floor made from microscopic shells
NOTES					

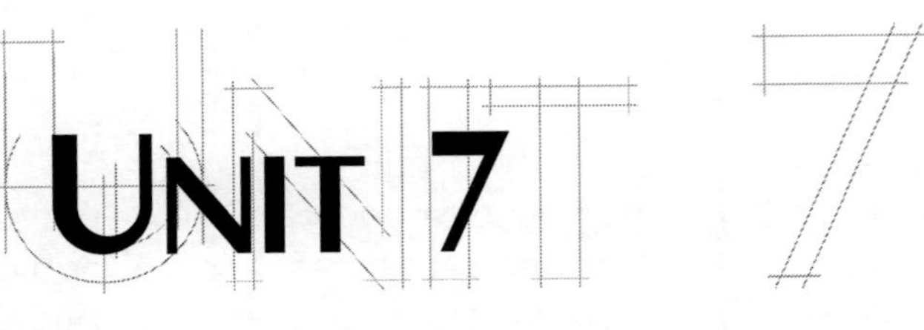

UNIT 7

Geologic Time

7

Geologic Time
Section 24

	Geologic Time	Relative Time	Absolute Time	Fossils I	Fossils II
5	The summary of Earth's history based on the rock record	Placing events in order without using actual dates	The time it takes a radioactive isotope to decay to half of its original amount	The remains or evidence of past life	Wood that has had all its original material replaced with minerals
10	Examples include the Paleozoic and Mesozoic.	The principle that states that processes occurring today on Earth have always occurred	The process by which radioactive isotopes emit tiny particles	The study of prehistoric life	A fossil that is a hollow space left in rock that forms when an organism dissolves
15	Examples include the Triassic and Jurassic.	The principle that states that older rock layers are on the bottom of a series of undisturbed layers	An isotope used to date once-living things up to about 70,000 years old	An example of an organism's hard parts that might become fossilized	A fossil that forms when minerals fill a hollow space left by a dissolved organism

20	Examples include the Pleistocene and Holocene.	The principle that states that a fault is younger than the rock it cuts across	The product of radioactive decay	A type of rock in which fossils mostly occur	Fossils that can be used to correlate or date rock layers
25	The longest segment of geologic time	The matching of rock from one area to another	A layer of sediment that is deposited on a yearly cycle	The indirect evidence of a past life, such as a burrow or footprint	A fossil that consists of a thin carbon imprint of an organism
BONUS	Known as the Age of Mammals	The principle that states that sedimentary rocks are deposited in nearly horizontal layers	The stable daughter isotope of uranium-238	The hardened resin from trees that encased insects	An area in southern California where pools of a sticky, dark substance has preserved ancient animals
NOTES					

7 Geologic Time
Section 25

	Precambrian Time	Paleozoic Era	Mesozoic Era	Cenozoic Era	The Rise of Humans
5	The exposed area of the oldest rocks of a continent	The Paleozoic is known as the age of this form of life.	The Mesozoic is known as the age of this form of life.	The Cenozoic is known as the age of this form of life.	The species to which humans belong
10	The Precambrian shield in North America	Great swamps in eastern North America and Europe formed this fossil fuel.	The breakup of this supercontinent begins	These mountains begin forming as India collides with Asia.	A distinguishing trait of humans meaning walking upright on two legs
15	Within 300 million years, the approximate age of Earth	Eastern mountains that started forming in the Ordovician period	A group of reptiles that became extinct probably when a meteorite struck Earth	A major fault in California formed during the Miocene epoch	Upright, bipedal primates, including humans

20	These events were probably responsible for the gases in the early atmosphere.	A combination of the Mississippian and Pennsylvanian periods	The word that describes the eating habits of theropod dinosaurs	An epoch that ended about 10,000 years ago when the ice sheets retreated from North America	Primates can grasp objects because of this feature on their hand.
25	The most common Precambrian fossils, these are mounds of cyanobacteria.	The most common fossils of the Cambrian period	The landscape became dominated by seed- bearing plants with flowers, called these.	The first of these plains animals, in the Miocene epoch, were the size of a large cat.	Hominid species that lived from 2 million to 1.5 million years ago and used simple tools
BONUS	The two eras that make up Precambrian time	The last period of the Paleozoic era	The last period of the Mesozoic era	The current epoch	The group of oldest known hominids
NOTES					

UNIT 8

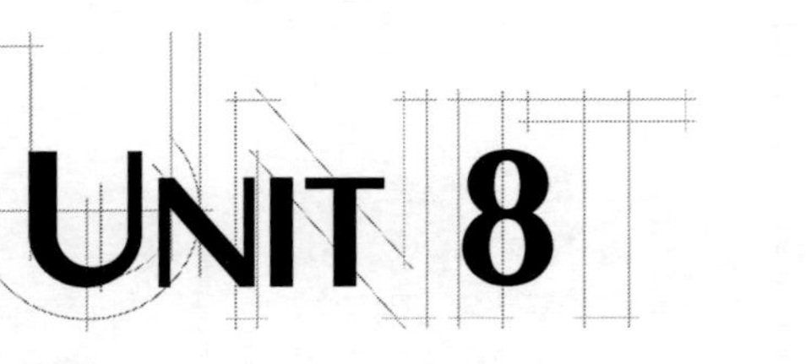

Resources and the Environment

8 Resources and the Environment

Section 26

	Resources	Land Resources	Air Resources	Water Resources	Resources and Population
5	All the resources that Earth provides	The place where an organism lives	A gas that our bodies need and that makes up 21% of air	A lake formed by the accumulation of water behind a dam	As a population increases, its demand for resources does this.
10	A resource that can be used indefinitely because it is replaced in nature	Cutting down all the trees in an area of forest	A gas in air that makes all rain slightly acidic	A long period of low rainfall	A pattern of growth in which a population grows quickly as it increases in size
15	A resource that exists in a fixed amount and is used up faster than it can be replaced	A layer of soil with the most organic matter in which crops grow	The process by which plants use carbon dioxide to make food	A permeable layer of rock that contains enough groundwater to supply wells	The number of organisms an environment can support

20	The known deposits of a mineral that is worth mining economically	The clearing of forests without replanting	When components are sufficient to keep a cycle going, the cycle is in a state of this.	The process of removing salts from ocean water to provide freshwater	Any factor that causes a population to stop growing
25	The replacement of renewable resources at the same rate at which they are used up	A process by which grassland becomes desert	A type of radiation that is mostly absorbed by ozone in the upper atmosphere	A pipe or passageway that brings water from a distance	A factor that affects a population more as the population density increases
BONUS	A resource used to make lubricants, plastics, fertilizers, and dyes	The natural accumulation of sand, gravel, and crushed stone	A gas produced by the radioactive decay of uranium that can accumulate to dangerous levels indoors	A large aquifer beneath the Great Plains	The current human population of the world, within 500 million
NOTES					

8 Resources and the Environment

Section 27

	Energy Use	Fossil Fuels	Alternative Sources I	Alternative Sources II	Conserving Energy
5	The ability to move an object or do work	The three main fossil fuels	Electricity produced by falling or fast-moving water	Heat from steam and hot water produced naturally within Earth	To collect and reuse materials
10	Energy sources that formed from organisms that lived millions of years ago	Gasoline and kerosene are derived from this fuel.	The process of splitting a nucleus to release energy	Plant material and fecal materials from animals that is burned to produce energy	The protection and management of natural resources
15	A machine that uses mechanical energy to produce electricity	A fossil fuel often found in deposits lying above oil	A type of solar heating system that includes the collection, storage, and distribution of heat	A mixture of gasoline and alcohol produced from grain crops	Energy can be conserved by replacing incandescent bulbs with these kind of bulbs.

20	Material that is burned to provide heat or power	The hardest, most efficient, and cleanest-burning type of coal	The element most often used to produce nuclear energy	This country gets most of its energy from geothermal sources along the mid-Atlantic ridge.	The use of energy resources in the most productive ways
25	The part of a generator that turns magnets near wires to produce electricity	The type of rock in which fossil fuels are found	A thin wafer that converts sunlight into electricity	Rock that contains a mixture of hydrocarbon compounds called kerogen	Using steam to produce both electricity and heat is an example of this process.
BONUS	A device that raises or lowers the voltage of electricity	Higher concentrations of this element make coal burn hotter and cleaner.	Within 5%, the percentage of electricity produced in the United States by hydroelectric power	The state with most of the wind turbines in the United States	Ratings that indicate the insulating abilities of construction materials
NOTES					

8 Resources and the Environment

Section 28

	Land Pollution	Air Pollution	Water Pollution	Pollution Solutions I	Pollution Solutions II
5	Any solid form of garbage	A hazy mixture of gases from pollutants reacting in sunlight	Waste that goes down the drain or is flushed down the toilet	Steplike cuts made in the side of a steep slope to slow soil erosion	A facility at which most harmful substances are removed from sewage
10	A method of solid-waste disposal in which garbage is left in an open area	Tiny, solid pieces of material that pollute the air	A source that produces water pollution from a single site	The process in which a mining company restores mined land to its original contours and replants vegetation	Yard wastes, food scraps, and other organic matter left to decompose and used as fertilizer
15	Chemicals applied to soil to increase crop yields	The increase in Earth's average surface temperature from excess CO_2 in the atmosphere	The adding of heated water to lakes or rivers	A federal agency in charge of enforcing environmental laws in the United States	A facility at which garbage is buried in such a way to minimize pollution

20	Chemicals applied to crops to kill crop-damaging pests	The major chemical in smog	A quick growth of algae from fertilizers that wash into a pond	A device on a smokestack that uses water to reduce sulfur-dioxide emissions	A farming method that decreases soil erosion by not plowing under crop wastes
25	Wastes that are toxic, explosive, infectious, or radioactive	It traps pollution in a layer of cooler air beneath a layer of warmer air.	A type of microscopic organism that pollutes water and can cause disease	Coal burns cleaner if it contain less of this element.	The main federal law that protects the nation's waters
BONUS	A process in which soil and rock is removed over a wide area to reach shallow deposits	Two kinds of chemicals that produce acid precipitation	These underground sewage-treatment containers sometimes leak and pollute groundwater.	The use of bacteria to consume toxic materials, such as spilled oil	A Great Lake that was cleaned of much of its pollution after being considered "dead" in the 1960s
NOTES					

UNIT 9

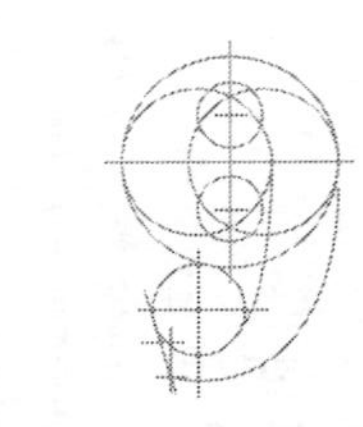

Astronomy

9 Astronomy

Section 29

	Radiation from Space	Telescopes	Satellites and Probes	Human Spaceflight I	Human Spaceflight II
5	The arrangement of different kinds of electromagnetic radiation from the Sun	A type of telescope that uses a lens to collect light	An object that revolves around another object in space	The United States federal agency in charge of spaceflight	The reusable spacecraft that have been launched into Earth's orbit since 1981
10	The part of the electromagnetic spectrum that humans can see	A type of telescope that uses a mirror to collect light	A spacecraft, without astronauts, sent to gather information about an object in space	A country that competed with the United States in trying to land astronauts on the moon	The space station that has been built by many nations
15	Unlike sound waves, electromagnetic waves can travel through this.	The point in a telescope at which incoming light comes together	The Hubble Space Telescope can take pictures without interference from this.	The first astronaut on the moon	Technologies developed for space programs that are now in common use

20	A type of radiation immediately to the right of violet light on the spectrum	A scientist who invented the first telescope to study outer space	The *Pathfinder* probe explored this planet.	The U.S. space program that landed astronauts on the moon	The two parts of the space shuttle that fall off after launch and are recovered for reuse
25	Radiation on the spectrum is arranged by wavelength and this property.	A telescope that detects the longest wavelengths of radiation	A type of orbit in which a satellite is always above the same point on Earth	The U.S. space program that sent the first U.S. astronauts into space	The first U.S. space station launched in 1973
BONUS	A type of radiation with the longest wavelengths	The process of linking separate telescopes together to act as one	The name of the first satellite ever launched	The U.S. space program that first practiced docking	The condition that exists when objects are falling freely together so that gravity seems to disappear
NOTES					

9 Astronomy

Section 30

	Earth in Space	Motions of the Moon	Moon Phases	Eclipses	Surface of the Moon
5	The spinning of Earth around its axis	The side of the Moon that always faces Earth	The phase when we see the whole lighted side of the Moon	It occurs when the Moon passes through Earth's shadow during full moon.	A round pit formed by the impact of an object
10	Earth does this almost exactly once around the Sun each year.	The Moon rises in this part of the sky.	The phase we see when the Sun lights the side of the Moon facing away from Earth	It occurs when the Moon passes directly between the Sun and Earth.	The dark, smooth plains on the Moon
15	The force that pulls Earth into orbit around the Sun	How long it takes the Moon to revolve around Earth	The phase in which half of the lighted side of the Moon is visible.	During a lunar eclipse, the Moon is still visible but often becomes this color.	Bright streaks of rocks that radiate out from a crater

20	The number of days in a leap year	The point at which the Moon is nearest to Earth	The increasing of the visible amount of the Moon's lighted surface after new moon	The darkest part of a shadow cast by Earth or the Moon	The layer of loose rock particles on the lunar surface
25	The shape of Earth's orbit	The point at which the Moon is farthest from Earth	The decreasing of the visible amount of the Moon's lighted surface after full moon	The area of partial shadow cast by Earth or the Moon	Trenchlike valleys in the Moon's maria
BONUS	The point at which Earth is nearest to the Sun	Within 10 minutes, how many minutes later the Moon rises each night or day	The phase in which almost all of the lighted side of the Moon is visible	It occurs when only part of the Moon is in Earth's umbra.	The type of rock that makes up most of the Moon's maria
NOTES					

9 Astronomy

Section 31

	Early Astronomers	Inner Planets	Outer Planets	Other Bodies in Space	Formation of Solar System
5	In the geocentric system of the ancient Greeks, all objects in space revolved around this.	The third planet from the sun	It has the most noticeable ring system made of pieces of rock and ice.	Bodies of rock, ice, and gas with varied orbits around the Sun	A force that causes interstellar material to come together enough to form stars
10	A Polish astronomer in the 1500s who developed a Sun-centered model of the solar system	The nearest planet to the Sun	The largest planet, characterized by the Great Red Spot	Most of these rocky bodies are located in a belt between Mars and Jupiter.	Most material in a rotating cloud of gas and dust gathers in the center, compresses, and forms this.
15	An Italian astronomer who discovered Jupiter's four largest moons	A lot of iron in the soil gives this planet a reddish color.	This outermost planet is smaller than Earth's moon.	A streak of light, also called a shooting star, made when a fragment passes through Earth's atmosphere	A cloud of gas and dust that forms stars and planets

20	A German mathematician who discovered that the orbit of each planet is an ellipse	Thick, yellowish clouds block our view of this planet's surface.	Instead of being rocky worlds, the four largest planets are made of these.	The part of a meteor that doesn't burn up in the atmosphere and strikes Earth's surface	Objects made of solid particles combining; ancestors of planets
25	An English scientist who explained why the planets stay in orbit	Its surface is heavily cratered, like Earth's moon.	The axis of this planet is tipped almost parallel to its orbital plane.	A cloud of gas that forms around the nucleus of a comet	Two main gases in interstellar clouds
BONUS	A Danish astronomer who made the most complete astronomical observations before the telescope was invented	Its southern polar ice cap may contain frozen CO_2.	Its largest moon, Triton, orbits backwards.	The farthest cluster of comets from the Sun	The name of the most widely accepted model of how our solar system formed
NOTES					

9 Astronomy

Section 32

	The Sun	Studying Stars	Characteristics of Stars	Star Life Cycles	Galaxies and the Universe
5	The combining of nuclei that gives the Sun its energy	A tool used to separate starlight into its colors	One of the 88 regions of stars in the sky	A cloud of gas and dust in which a star forms	A group of millions or billions of stars held together by gravity
10	A dark spot on the surface of the Sun	A spectrum made of an unbroken band of colors	A measure of how bright a star appears from Earth	The main sequence stars are actively fusing hydrogen into this element.	The galaxy to which our solar system belongs
15	Colorful displays resulting from the solar wind colliding with gases in Earth's atmosphere	The shift toward the blue end of the spectrum that means a star is moving toward Earth	The actual brightness of a star	The remnant of a massive star, with gravity so intense that not even light can escape	The theory that says the universe began from a small area in space and has been expanding ever since

20	An eruption of particles and radiation from the surface of the Sun	A spectrum of unevenly spaced lines of different colors and brightnesses	The color of the hottest stars	The explosion that follows the collapse of the core of a massive star	A type of galaxy that resembles a pinwheel
25	The top layer of the Sun's atmosphere	A continuous spectrum with dark lines crossing it	A measure of how bright a star would be if all stars were the same distance from Earth	The glowing core of what used to be a red giant	Extremely bright and distant objects whose name is short for *quasi-stellar radio sources*
BONUS	The length of time of the sunspot cycle	The apparent change in wavelength of radiation that lets astronomers tell how a star is moving	A unit of measure in space, which is the average distance between Earth and the Sun	A diagram that plots the luminosity of stars against their surface temperatures	The highly energetic cores of some galaxies
NOTES					

Answer Key

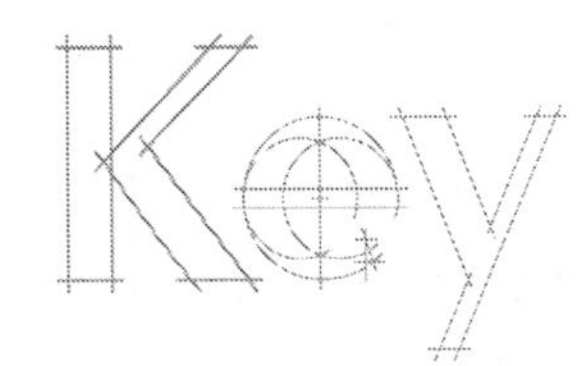

1 Introductory Material
Section 1

	Nature of Earth Science	Earth's Systems	Scientific Methods	Measurement I	Measurement II
5	What is astronomy?	What is the atmosphere?	What is identifying the problem?	What is a meter?	What is a kilometer?
10	What is oceanography?	What is the hydrosphere?	What is a hypothesis?	What is mass?	What is 10?
15	What is meteorology?	What is the biosphere?	What is a control?	What is weight?	What is Celsius?
20	What is geology?	What is the lithosphere?	What is an independent variable?	What is volume?	What is area?
25	What is technology?	What is the asthenosphere?	What is a theory?	What is density?	What is Kelvin?
BONUS	What is hydrology?	What two gases make up most of Earth's atmosphere?	What is a scientific law?	What is the cubic meter?	What is the International System of Units (or Systeme International d'Unites)?

1 Introductory Material
Section 2

	Models	Parts of a Map	Types of and Projections of Maps	Topographic Maps I	Topographic Maps II
5	What is a model?	What are symbols?	What is a topographic map?	What is a contour line?	What is large scale?
10	What is a globe?	What is a key or legend?	What is a geologic map?	What is the contour interval?	What is a benchmark?
15	What is a map?	What is a scale?	What is a Mercator projection?	What is a depression?	What is a hill or mountain?
20	What is a scale model?	What is a compass rose or north arrow?	What is a conic projection?	What is the U. S. Geological Survey, or USGS?	What is upstream?
25	What is a mental model?	What is detail?	What is a gnomonic or planar projection?	What is heavier and darker on a topographic map?	What is a cliff or steep slope?
BONUS	What is the solar system?	What is 1:63,500 or 1/63,500?	What are cartographers?	What is pink?	What is 100?

1 Introductory Material
Section 3

	Latitude and Longitude	Time Zones	Remote Sensing	Continents	Landforms
5	What is latitude?	What is east, or west to east?	What are satellites?	What is seven?	What are mountains?
10	What is longitude?	What is 24?	What is Landsat?	What is Eurasia?	What are valleys?
15	What is the Northern Hemisphere?	What is subtract an hour or gets one hour earlier in each time zone?	What is the Global Positioning System (GPS)?	What is Australia?	What are plains?
20	What are minutes?	What is the International Date Line?	What is sonar?	What is Antarctica?	What are plateaus?
25	What is 90°S?	What is noon?	What are false-color images?	What are North America and South America?	What are coastal plains?
BONUS	What is the prime meridian?	What is 15?	What is a receiver?	What are South America and Africa?	What is relief?

2 Earth Materials
Section 4

	Atoms	Phases of Matter	Elements	Combinations of Matter	Energy
5	What is an element?	What is a solid?	What are protons?	What is a compound?	What is kinetic energy?
10	What is an atom?	What is a liquid?	What is atomic number?	What is a covalent bond?	What is potential energy?
15	What is a proton?	What is evaporation?	What are isotopes?	What is a solution?	What is nuclear energy or atomic energy?
20	What is a neutron?	What is condensation?	What is the periodic table of elements?	What is a molecule?	What are hydrogen atoms?
25	What is an ion?	What is plasma?	What is mass number?	What is an alloy?	What is the law of conservation of energy?
BONUS	What is 92?	What is sublimation?	What is oxygen?	What is water?	What are chemical bonds?

2 Earth Materials
Section 5

	Mineral Characteristics	Mineral Groups	Identifying Minerals I	Identifying Minerals II	Mineral Resources
5	What is inorganic?	What are silicates?	What is color?	What is cleavage?	What is an ore?
10	What is naturally?	What is quartz?	What is luster?	What is fracture?	What are gems?
15	What is solid?	What is feldspar?	What is texture?	What is carbon dioxide?	What is a vein?
20	What is a crystal?	What are carbonates?	What is streak?	What is double refraction?	What is an open pit mine?
25	What is magma?	What are oxides?	What is hardness?	What is specific gravity?	What is bauxite?
BONUS	What is 3,000?	What is the silica tetrahedron?	What is diamond?	What is magnetite?	What is amethyst?

2 Earth Materials
Section 6

	Introduction to Rocks	Formation of Igneous Rocks	Classifying Igneous Rocks	Formation of Metamorphic Rocks	Classifying Metamorphic Rocks
5	What are minerals?	What is magma?	What is mineral composition?	What is parent rock?	What are foliated and nonfoliated?
10	What is the rock cycle?	What is lava?	What is granite?	What is regional metamorphism?	What is foliated?
15	What is igneous rock?	What are intrusive rocks?	What is pumice?	What is contact metamorphism?	What is slate?
20	What is sedimentary rock?	What are extrusive rocks?	What is basalt?	What is deformational metamorphism?	What is marble?
25	What is metamorphic rock?	What is glassy?	What is a porphyry?	What is hydrothermal metamorphism?	What is gneiss?
BONUS	What is *fire?*	What is mafic?	What is rhyolite?	What are porphyroblasts?	What is quartzite?

2 Earth Materials
Section 7

	Formation of Sedimentary Rocks	Classifying Sedimentary Rocks I	Classifying Sedimentary Rocks II	Features of Sedimentary Rocks	Name That Rock
5	What are sediments?	What is clastic rock?	What is a conglomerate?	What are fossils?	What is granite?
10	What is deposition?	What is organic rock?	What is shale?	What is bedding?	What is obsidian?
15	What is compaction?	What is chemical rock?	What is chalk?	What are ripple marks?	What is limestone?
20	What is cementation?	What is limestone?	What is sandstone?	What is cross-bedding?	What is slate?
25	What is lithification?	What is coarse-grained?	What is limestone?	What is a geode?	What is breccia?
BONUS	What is gray or white?	What is coal?	What is coquina?	What is a concretion?	What is schist?

3 Earth's Internal Processes
Section 8

	Structure of Earth	Drifting Continents	Seafloor Spreading	Plate Tectonics I	Plate Tectonics II
5	What is the crust?	What is continental drift?	What is the Mid-Atlantic Ridge?	What is a plate?	What are convection currents?
10	What is the mantle?	What is Pangaea?	What are trenches?	What is a convergent or colliding boundary?	What is lower density?
15	What is the inner core?	What is South America?	What is younger?	What is a divergent or spreading boundary?	What is ridge push?
20	What is the outer core?	What are fossils?	What is ocean crust?	What is a transform boundary?	What is slab pull?
25	What is the lithosphere?	What is how the continents move?	What is Earth's magnetic field?	What is subduction?	What is Gondwana?
BONUS	What is the asthenosphere?	Who is Alfred Wegener?	Who is Harry Hess?	What is 2 to 3 cm/y?	What is the Great Rift Valley?

3 Earth's Internal Processes
Section 9

	Faults	Earthquake Waves	Measuring and Locating Earthquakes	Earthquake Hazards	Earthquake Prediction
5	What is a fault?	What are seismic waves?	What is the focus?	What is a shallow focus?	What is Alaska, California, or Hawaii?
10	What is a strike-slip fault?	What are primary, or P, waves?	What is the epicenter?	What is a tsunami?	What is a fault?
15	What is the hanging wall?	What are secondary, or S, waves?	What is a seismograph?	What are aftershocks?	What are plate boundaries?
20	What is a reverse fault?	What are surface waves?	What is the Richter scale?	What is fire?	What is a seismic gap?
25	What is a thrust fault?	What are surface waves?	What is three?	What is liquefaction?	What is stress?
BONUS	What is a fault scarp?	What are body waves?	What is the moment magnitude scale?	What is the modified Mercalli scale?	What is magnitude?

3 Earth's Internal Processes
Section 10

	Magma	Volcanic Landforms	Volcanoes and Plate Tectonics	Intrusive Features	Volcanic Eruptions
5	What is lava?	What is a vent?	What are plate boundaries?	What is a pluton?	What is Mount St. Helens?
10	What is viscosity?	What is a cinder cone?	What is a volcanic island arc?	What is a dike?	What is a pyroclastic flow?
15	What is basaltic magma?	What is a shield volcano?	What is a hot spot?	What are batholiths?	What is pahoehoe?
20	What is rhyolitic magma?	What is a composite volcano?	What is Iceland?	What is a dike?	What is pillow lava?
25	What causes magma to form?	What is a lava plateau?	What is a rift zone?	What is a stock?	What is a tiltmeter?
BONUS	What is andesitic magma?	What is a caldera?	What is a change in direction of the Pacific Plate?	What is an atoll?	What is tephra?

3 Earth's Internal Processes
Section 11

	Isostasy	Convergent Boundary Mountains I	Convergent Boundary Mountains II	Divergent and Nonboundary Mountains	Mountain Ranges
5	What is density?	What are plates?	What are continental plates?	What is a volcano?	What are the Himalayas?
10	What is buoyancy?	What are folded mountains?	What are oceanic plates?	What are uplifted or dome mountains?	What are the Appalachians?
15	What is the principle of isostasy?	What is a syncline?	What are oceanic and continental plates?	What are fault-block mountains?	What is erosion?
20	What is oceanic?	What is an anticline?	What are limbs?	What are ocean ridges?	What are the Black Hills?
25	What is isostatic rebound?	What is compression?	What is orogeny?	What is tension?	What are the Andes?
BONUS	What are roots?	What is strike?	What is dip?	What is a horst?	What is the Teton Range?

4 Earth's Changing Surface
Section 12

	Mechanical Weathering	Chemical Weathering	Weathering Rates	Soil Formation	Soil Types
5	What is weathering?	What is rust?	What is warm and moist?	What is humus?	What are the climates in which they form?
10	What is its composition?	What is hydrolysis?	What is surface area?	What is weathered rock?	What is clay?
15	What are roots?	What is oxidation?	What is sedimentary?	What is a soil profile?	What is loam?
20	What is frost wedging?	What is carbonic acid?	What is quartz?	What is a soil horizon?	What is arctic soil?
25	What is exfoliation?	What is calcite?	What is polar?	What is the B horizon?	What is temperate soil?
BONUS	What is burrow?	What is kaolinite?	What is hilly or mountainous?	What is the C horizon?	What is desert soil?

4 Earth's Changing Surface
Section 13

	River Systems	River Valleys	River Erosion	River Deposition	Floods and Floodplains
5	What is downhill or downslope?	What is V-shaped?	What is erosion?	What is deposition?	What is a floodplain?
10	What is a tributary?	What is a canyon or gorge?	What is abrasion?	What is a delta?	What is an oxbow lake?
15	What is a drainage basin or watershed?	What is a gully?	What is the outside of the meander?	What is the inside of the meander?	What is sideways or laterally?
20	What is gradient?	What is a meander?	What is headward erosion?	What is an alluvial fan?	What is a levee?
25	What is a divide?	What is the mature stage?	What is load?	What are distributaries?	What is a flash flood?
BONUS	What is dendritic?	What is base level?	What is turbulence?	What is New Orleans?	What is a yazoo stream?

4 Earth's Changing Surface
Section 14

	Groundwater	Groundwater Erosion and Deposition	Mass Movements	Wind Erosion	Wind Deposition
5	What is the water table?	What is dissolves?	What is gravity?	What is a dust storm?	What is a sand dune?
10	What is a spring?	What is a sinkhole?	What is a mudflow?	What is arid or dry?	What is loess?
15	What is the zone of saturation?	What is karst topography?	What is creep?	What is deflation?	What is the windward side?
20	What is permeability?	What is a stalactite?	What is a landslide?	What is desert pavement?	What is the leeward side?
25	What is the zone of aeration?	What is a stalagmite?	What is slump?	What is saltation?	What is quartz?
BONUS	What is capillary action?	What is hard water?	What is talus?	What are ventifacts?	What is dune migration?

4 Earth's Changing Surface
Section 15

	Glaciers	Glacial Movement	Glacial Erosion	Glacial Deposition	Ice Ages
5	What is ice?	What is gravity?	What is U-shaped?	What is till?	What are continental glaciers?
10	What is a valley glacier?	What is friction?	What is a cirque?	What is a moraine?	What are the Great Lakes?
15	What is a continental glacier?	What are icebergs?	What is a kettle lake?	What are drumlins?	What is a woolly mammoth?
20	What is the snow line?	What are crevasses?	What is a horn?	What is outwash?	What is about 11,000?
25	What is Greenland or Antarctica?	What is the surface or center?	What is a fjord?	What is an esker?	What is an interglacial period?
BONUS	What is firn?	What is basal slip?	What are striations?	What is an erratic?	What are ocean currents?

4 Earth's Changing Surface
Section 16

	Waves	Shoreline Waves and Currents	Shoreline Erosion	Shoreline Deposition I	Shoreline Deposition II
5	What is a crest?	What is a breaker?	What is a sea cave?	What is a beach?	What is a barrier island?
10	What is a trough?	What is wave height?	What is a sea arch?	What are rivers?	What are plants?
15	What is wave height?	What is backwash?	What are headlands?	What is longshore drift?	What is a hook?
20	What is wavelength?	What is a longshore current?	What is a sea stack?	What is a spit?	What is a groin?
25	What is period?	What is refraction?	What is abrasion?	What is a baymouth bar?	What is a breakwater?
BONUS	What is fetch?	What is a rip current?	What is a wave-cut platform?	What is a lagoon?	What is a tombolo?

5 The Atmosphere
Section 17

	Composition of the Atmosphere	Structure of the Atmosphere	Heat in the Atmosphere	Air Pressure	Wind
5	What is oxygen?	What is the troposphere?	What is the sun?	What is decreases?	What is low pressure?
10	What is water vapor?	What is decreases?	What is radiation?	What is a barometer?	What are prevailing winds?
15	What is salt?	What is the stratosphere?	What is conduction?	What is increases?	What is a sea breeze?
20	What is ozone?	What is the stratopause?	What is convection?	What is an isobar?	What are jet streams?
25	What is carbon dioxide?	What is the ionosphere?	What is temperature?	What is the millibar?	What is the Coriolis effect?
BONUS	What is 21%?	What is the troposphere?	What is heat?	Who is Evangelista Torricelli?	What is the intertropical convergence zone (ITCZ)?

5 The Atmosphere
Section 18

	Water Cycle	Humidity	Cloud Formation	Cloud Types	Precipitation
5	What is the atmosphere?	What is water vapor?	What is the dew point?	What is fog?	What is collide?
10	What is evaporation?	What is saturated?	What is condensation nuclei?	What are cumulus clouds?	What is a rain gauge?
15	What is condensation?	What is relative humidity?	What is move farther apart?	What are cirrus clouds?	What is hail?
20	What is precipitation?	What is a psychrometer?	What is latent heat?	What are stratus clouds?	What is freezing rain?
25	What is transpiration?	What is 55%?	What are ice crystals?	What are cumulonimbus clouds?	What is sleet?
BONUS	What is runoff?	What is wet-bulb and dry-bulb?	What is stability?	What is alto–?	What is coalescence?

5 The Atmosphere
Section 19

	Air Masses	Fronts	Pressure Systems	Collecting Weather Data	Forecasting Weather
5	What are temperature and humidity?	What are air masses?	What is rotate?	What is a weather balloon?	What is a meteorologist?
10	What is continental polar?	What is a cold front?	What is clockwise?	What is Doppler radar?	What is rainy or stormy weather?
15	What are warm and moist?	What is a warm front?	What is counterclockwise?	What is an anemometer?	What is fair weather?
20	What is maritime polar?	What is a stationary front?	What is a low-pressure system?	What is temperature?	What are weather models?
25	What is continental tropical?	What is an occluded front?	What is a polar front?	What is a radiosonde?	What is the National Weather Service?
BONUS	What is air mass modification?	What are cirrus clouds?	What is upper-air flow?	What is the Automated Surface Observing System (ASOS)?	What is an analog forecast?

5 The Atmosphere
Section 20

	Thunderstorms	Tornadoes	Hurricanes	Other Extreme Weather	Weather Maps
5	What is thunder?	What is a low?	What are the tropics?	What is a blizzard?	What is a high or high-pressure area?
10	What is the cumulus stage?	What is direction?	What is warm, moist air?	What is a drought?	What is a cold front?
15	What are downdrafts?	What is spring and early summer?	What is the eye?	What is the wind-chill factor?	What are isobars?
20	What is a supercell?	What is a vortex?	What is a storm surge?	What is a flash flood?	What are isotherms?
25	What are frontal thunderstorms?	What is a tornado watch?	What is the eyewall?	What is the heat index?	What is a station model?
BONUS	What is negative?	Who is Tetsuya "Ted" Fujita?	What is 120 km/h (74 mph)?	What is a nor'easter?	What is dew point?

5 The Atmosphere
Section 21

	Climate	Climate Classification	Natural Climate Changes	Humans Affect Climates	Seasons
5	What are temperature and precipitation?	What are subclimates?	What is sunlight?	What is global warming?	What is the tilt of Earth's axis?
10	What is the leeward side?	What is tundra?	What is El Niño?	What is a climatologist?	What is winter?
15	What is a sphere?	What is semiarid?	What is Earth's axis?	What is burning fossil fuels?	What is the summer solstice?
20	What is the daily temperature range?	What is a microclimate?	What are sunspots?	What is the greenhouse effect?	What is the vernal equinox?
25	What are heat up and cool down?	What is a heat island?	What is carbon dioxide?	What is deforestation?	What is 12?
BONUS	What is the Gulf Stream?	What is the Koeppen Classification System?	What is an ice core?	What is infrared radiation?	What is the Tropic of Cancer?

6 The Oceans
Section 22

	Oceans of the World	Properties of Seawater	Structure of the Oceans	The Ocean Floor I	The Ocean Floor II
5	What is oceanography?	What is salinity?	What is dark?	What is a continental shelf?	What are mid-ocean ridges?
10	What is the Pacific Ocean?	What is sodium chloride (NaCl)?	What is decreases?	What is a continental slope?	What are abyssal hills?
15	What is the Arctic Ocean?	What is evaporation?	What is the mixed layer?	What are deep-sea trenches?	What is a submarine canyon?
20	What is echo sounding or sonar?	What is seawater?	What is the thermocline?	What is the abyssal plain?	What is a deep-sea vent?
25	What is 71%?	What is fresh water?	What is the mixed layer?	What is a seamount?	What is a continental rise?
BONUS	What is the H.M.S. *Challenger?*	What is 35 parts per thousand or 3.5%?	What is polar water?	What are guyots?	What is a fracture zone?

6 The Oceans
Section 23

	Surface Currents	Underwater Currents	Tides	Ocean Life	Ocean Resources
5	What is wind?	What is a density current?	What is gravity?	What are phytoplankton?	What is salt?
10	What is the equator?	What is colder?	What is the Moon?	What are nekton?	What is desalination?
15	What is a countercurrent?	What is upwelling?	What is a neap tide?	What are zooplankton?	What is kelp?
20	What is counterclockwise?	What is a turbidity current?	What is a spring tide?	What are corals?	What is overfishing?
25	What are the trades and westerlies?	What are nutrients?	What is the tidal range?	What is chemosynthesis?	What is manganese?
BONUS	What is a gyre?	What is the west?	What is the Bay of Fundy, Canada?	What is hydrogen sulfide?	What are oozes?

7 Geologic Time
Section 24

	Geologic Time	Relative Time	Absolute Time	Fossils I	Fossils II
5	What is the geologic time scale?	What is relative dating?	What is half-life?	What are fossils?	What is petrified wood?
10	What are eras?	What is uniformitarianism?	What is radioactive decay?	What is paleontology?	What is a mold?
15	What are periods?	What is superposition?	What is carbon-14?	What is tooth, bone, or shell?	What is a cast?
20	What are epochs?	What is cross-cutting relationships?	What is a daughter isotope?	What is sedimentary rock?	What are index fossils?
25	What is an eon?	What is correlation?	What is a varve?	What is a trace fossil?	What is a carbonaceous film?
BONUS	What is the Cenozoic era?	What is original horizontality?	What is lead-206?	What is amber?	What are the La Brea Tar Pits?

7 Geologic Time
Section 25

	Precambrian Time	Paleozoic Era	Mesozoic Era	Cenozoic Era	The Rise of Humans
5	What is a shield?	What are invertebrates?	What are reptiles?	What are mammals?	What is *Homo sapiens?*
10	What is the Canadian shield?	What is coal?	What is Pangaea?	What are the Himalayas?	What is bipedal?
15	What is 4.6 billion years?	What are the Appalachians?	What are dinosaurs?	What is the San Andreas Fault?	What are hominids?
20	What are volcanic eruptions?	What is the Carboniferous period?	What is carnivorous?	What is the Pleistocene?	What is an opposable thumb?
25	What are stromatolites?	What are trilobites?	What are angiosperms?	What are horses?	What is *Homo habilis?*
BONUS	What are the Proterozoic and Archean eras?	What is the Permian period?	What is the Cretaceous period?	What is the Holocene or Recent?	What is *Australopithecus?*

8 Resources and the Environment
Section 26

	Resources	Land Resources	Air Resources	Water Resources	Resources and Population
5	What are natural resources?	What is a habitat?	What is oxygen?	What is a reservoir?	What is increases?
10	What is a renewable resource?	What is clearcutting?	What is carbon dioxide?	What is drought?	What is exponential growth?
15	What is a nonrenewable resource?	What is topsoil?	What is photosynthesis?	What is an aquifer?	What is carrying capacity?
20	What is a reserve?	What is deforestation?	What is balance or equilibrium?	What is desalination?	What is a limiting factor?
25	What is sustainable yield?	What is desertification?	What is ultraviolet radiation?	What is an aqueduct?	What is a density-dependent factor?
BONUS	What is petroleum?	What is aggregate?	What is radon?	What is the Ogallala Aquifer?	What is 6 billion?

8 Resources and the Environment
Section 27

	Energy Use	Fossil Fuels	Alternative Sources I	Alternative Sources II	Conserving Energy
5	What is energy?	What are coal, oil or petroleum, and natural gas?	What is hydroelectricity?	What is geothermal energy?	What is recycle?
10	What are fossil fuels?	What is oil or petroleum?	What is nuclear fission?	What is biomass?	What is conservation?
15	What is a generator?	What is natural gas?	What is active solar heating?	What is gasohol?	What are fluorescent bulbs?
20	What is fuel?	What is anthracite?	What is uranium?	What is Iceland?	What is energy efficiency?
25	What is a turbine?	What is sedimentary rock?	What is a photovoltaic cell?	What is oil shale?	What is cogeneration?
BONUS	What is a transformer?	What is carbon?	What is approximately 10%?	What is California?	What are R-values?

8 Resources and the Environment
Section 28

	Land Pollution	Air Pollution	Water Pollution	Pollution Solutions I	Pollution Solutions II
5	What is solid waste?	What is smog?	What is sewage?	What are terraces?	What is a sewage-treatment plant?
10	What is open dumping?	What are particulates?	What is a point source?	What is reclamation?	What is compost?
15	What are fertilizers?	What is global warming?	What is thermal pollution?	What is the Environmental Protection Agency (EPA)?	What is a sanitary landfill?
20	What are pesticides?	What is ozone?	What is an algae bloom?	What is a wet scrubber?	What is no-till farming?
25	What are hazardous wastes?	What is a temperature inversion?	What are bacteria or viruses?	What is sulfur?	What is the Clean Water Act?
BONUS	What is strip mining?	What are sulfur dioxide and nitrogen oxides?	What are septic tanks?	What is bioremediation?	What is Lake Erie?

9 Astronomy
Section 29

	Radiation from Space	Telescopes	Satellites and Probes	Human Spaceflight I	Human Spaceflight II
5	What is the electromagnetic spectrum?	What is a refracting telescope?	What is a satellite?	What is the National Aeronautics and Space Administration (NASA)?	What are space shuttles?
10	What is visible light?	What is a reflecting telescope?	What is a space probe?	What is the Soviet Union?	What is the International Space Station (ISS)?
15	What is empty space?	What is the focus or focal point?	What is Earth's atmosphere?	Who is Neil Armstrong?	What are spinoffs?
20	What is ultraviolet radiation?	Who is Galileo Galilei?	What is Mars?	What is *Apollo?*	What are the booster rockets?
25	What is frequency?	What is a radio telescope?	What is a geosynchronous orbit?	What is *Mercury?*	What is *Skylab?*
BONUS	What are radio waves?	What is interferometry?	What is *Sputnik I?*	What is *Gemini?*	What is microgravity?

9 Astronomy
Section 30

	Earth in Space	Motions of the Moon	Moon Phases	Eclipses	Surface of the Moon
5	What is rotation?	What is the near side?	What is a full moon?	What is a lunar eclipse?	What is a crater?
10	What is revolves?	What is the east?	What is a new moon?	What is a solar eclipse?	What are maria?
15	What is gravity?	What is $27\frac{1}{3}$ days?	What is first or third quarter?	What is red or copper-colored?	What are rays?
20	What is 366?	What is perigee?	What is waxing?	What is the umbra?	What is regolith?
25	What is an ellipse?	What is apogee?	What is waning?	What is the penumbra?	What are rilles?
BONUS	What is perihelion?	What is 50 minutes?	What is waxing gibbous or waning gibbous?	What is a partial lunar eclipse?	What is basalt?

9 Astronomy
Section 31

	Early Astronomers	Inner Planets	Outer Planets	Other Bodies in Space	Formation of Solar System
5	What is Earth?	What is Earth?	What is Saturn?	What are comets?	What is gravity?
10	Who is Nicolaus Copernicus?	What is Mercury?	What is Jupiter?	What are asteroids?	What is a star?
15	Who is Galileo Galilei?	What is Mars?	What is Pluto?	What is a meteor?	What is an interstellar cloud?
20	Who is Johannes Kepler?	What is Venus?	What are gases?	What is a meteorite?	What are planetesimals?
25	Who is Isaac Newton?	What is Mercury?	What is Uranus?	What is the coma?	What are hydrogen and helium?
BONUS	Who is Tycho Brahe?	What is Mars?	What is Neptune?	What is the Oort cloud?	What is the nebular hypothesis?

9 Astronomy
Section 32

	The Sun	Studying Stars	Characteristics of Stars	Star Life Cycles	Galaxies and the Universe
5	What is fusion?	What is a spectroscope?	What is a constellation?	What is a nebula?	What is a galaxy?
10	What is a sunspot?	What is a continuous spectrum?	What is apparent magnitude?	What is helium?	What is the Milky Way?
15	What are auroras?	What is a blueshift?	What is luminosity?	What is a black hole?	What is the Big Bang theory?
20	What is a solar flare?	What is an emission spectrum?	What is bluish white?	What is a supernova?	What is a spiral galaxy?
25	What is the corona?	What is an absorption spectrum?	What is absolute magnitude?	What is a white dwarf?	What are quasars?
BONUS	What is 11 years?	What is the Doppler effect?	What is an astronomical unit (AU)?	What is the Hertzsprung-Russell (H-R) diagram?	What are active galactic nuclei (AGNs)?

Share Your Bright Ideas

We want to hear from you!

Your name____Date_________________________

School name____

School address____

City __________________________State _____Zip__________Phone number (______)__________

Grade level(s) taught________Subject area(s) taught____

Where did you purchase this publication?____

In what month do you purchase a majority of your supplements?_____________________________

What moneys were used to purchase this product?

___School supplemental budget ___Federal/state funding ___Personal

Please "grade" this Walch publication in the following areas:

Quality of service you received when purchasing .... A B C D
Ease of use .... A B C D
Quality of content .... A B C D
Page layout .... A B C D
Organization of material .... A B C D
Suitability for grade level .... A B C D
Instructional value .... A B C D

COMMENTS:____

____

What specific supplemental materials would help you meet your current—or future—instructional needs?

____

Have you used other Walch publications? If so, which ones?_______________________________

May we use your comments in upcoming communications? ___Yes ___No

Please **FAX** this completed form to **888-991-5755**, or mail it to

Customer Service, Walch Publishing, P. O. Box 658, Portland, ME 04104-0658

We will send you a **FREE GIFT** in appreciation of your feedback. **THANK YOU!**